U.S. SEISMIC ZONES

THE SAN ANDREAS FAULT

GREG ROZA

PowerKiDS press

Published in 2026 by The Rosen Publishing Group, Inc.
2544 Clinton Street, Buffalo, NY 14224

First Edition

Editor: Greg Roza
Book Design: Michael Flynn

Photo Credits: Cover Ningaloo.gg/Shutterstock.com; p. 5 Yarr65/Shutterstock.com; p. 6 Designua/Shutterstock.com; p. 7 VectorMine/Shutterstock.com; pp. 8, 10 Barks/Shutterstock.com; p. 9 https://commons.wikimedia.org/wiki/File:Falla_normal_Morro_Solar_Peru.jpg; p. 11 https://en.wikipedia.org/wiki/File:South_Island_blizzard_2003.jpg; p. 13 Zaporizhzhia vector/Shutterstock.com; p. 15 Jonas D Bell/Shutterstock.com; pp. 17, 19, 20 Everett Collection/Shutterstock.com; p. 21 https://commons.wikimedia.org/wiki/File:Mayor_Eugene_E._Schmitz.jpg; p. 23 Jon Leonoudakis / Alamy Stock Photo; p. 24 https://commons.wikimedia.org/wiki/File:NimitzFreewayRibbonLomaPrietaQuake.jpg; p. 25 https://commons.wikimedia.org/wiki/File:Bay_Bridge_Collapse_-_Flickr_-_sanbeiji.jpg; p. 27 Christopher E Brown; p. 29 Joel Hensler/Shutterstock.com.

Cataloging-in-Publication Data

Names: Roza, Greg.
Title: The San Andreas Fault / Greg Roza.
Description: Buffalo, New York : PowerKids Press, 2026. | Series: U.S. seismic zones | Includes glossary and index.
Identifiers: ISBN 9781499453416 (pbk.) | ISBN 9781499453423 (library bound) | ISBN 9781499453430 (ebook)
Subjects: LCSH: Earthquakes--Juvenile literature. | Earthquakes--California--Juvenile literature. | Seismology--Juvenile literature.
Classification: LCC QE521.3 R69 2026 | DDC 551.2209--dc23

Manufactured in the United States of America

CPSIA Compliance Information: Batch #CSPK26. For further information contact Rosen Publishing at 1-800-237-9932.

CONTENTS

BIG STATE, BIG QUAKES!

The large state of California is a land of opposites. It has snow-capped mountains, green valleys, beautiful coastlines, and blazing deserts. It features fertile farmland and densely packed urban areas. In the north, the world's largest trees, giant sequoias, tower over the Sierra Nevada mountains. In the south, world-famous beaches line the coast.

California also experiences two-thirds of all earthquakes that occur in the United States, making it a **significant** seismic zone. Seismic activity in California is the result of the many faults that cross the state. Much of this seismic activity takes place around the San Andreas Fault.

Scientists have long studied the geography of California in hopes of understanding the area's seismic activity. This has prepared the people of the state for future quakes. However, California has been struck by incredibly powerful earthquakes that are hard to prepare for. Scientists say the chances are good that the state will experience more of them in the future.

Large, destructive earthquakes hit California in 1906 and 1989. Many, many other earthquakes have hit the state. In fact, dozens of quakes hit the area every day, although most are too weak to feel.

IN THE ZONE

Seismic zones are areas where earthquakes are common. Earthquakes are caused by powerful forces at work inside Earth, and most occur near fault lines. A fault line is a crack that forms between or on massive pieces of Earth's upper layer called plates. The study of Earth's plates and their movements is called plate tectonics.

Scientists have discovered that Earth's plates are part of a layer called the lithosphere. This layer "floats" on another more molten layer called the aesthenosphere. At one time long ago, all of Earth's landmasses were packed together in a single landmass that scientists call Pangea. Over time, faults appeared in the land, and massive sections of the lithosphere drifted away from each other. In time, this action created the world and it's seven continents that we are familiar with today.

PANGEA

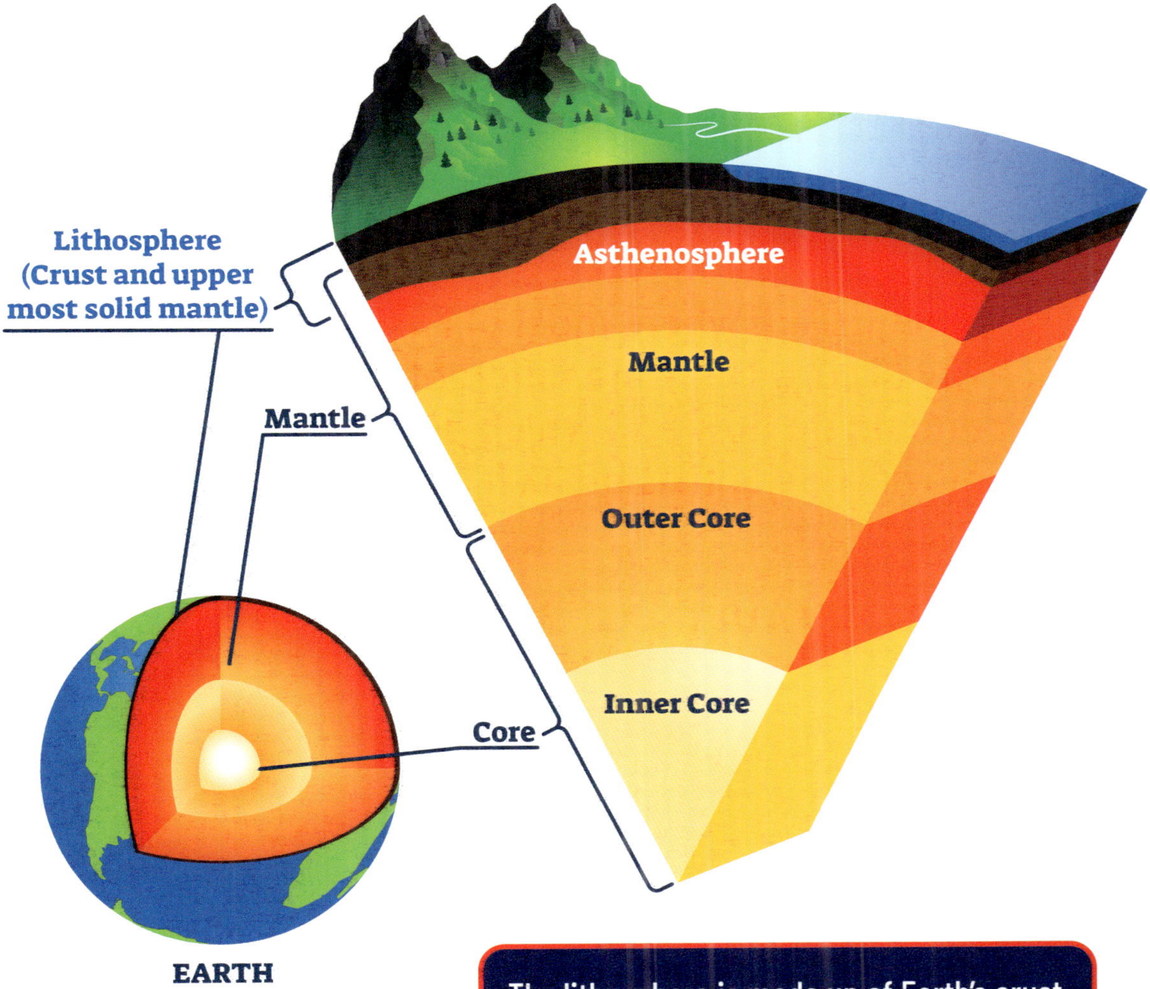

The lithosphere is made up of Earth's crust and the top layer of Earth's mantle.

EARTH-SHAKING FACT!

In 1912, German scientist Alfred Wegener described his theory of "continental drift" to the scientific world. His concepts remained controversial until the late 1960s. Today, plate tectonics and Wegener's theory of continental drift are widely held to be factual.

TYPES OF FAULTS

Due to continental drift, Earth's plates are constantly on the move. The spaces where plates meet, faults, are where most of Earth's seismic activity occurs. There are different types of faults depending on how and which direction the plates are moving. Scientists break faults down into 3 main kinds.

NORMAL FAULT

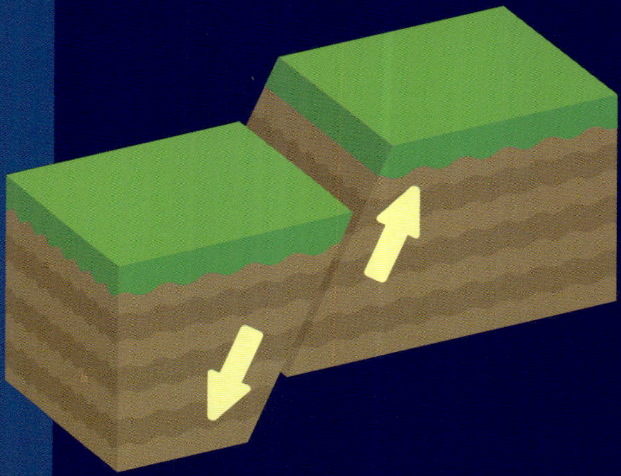

REVERSE FAULT

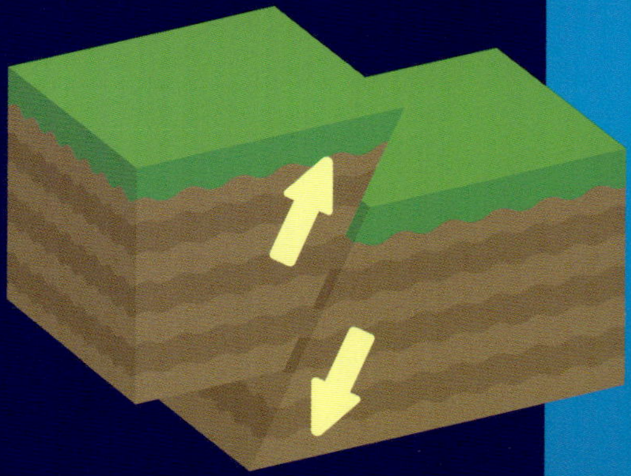

EARTH-SHAKING FACT!

Both normal and reverse faults, which are also called dip-slip faults, result in seismic activity, volcanic activity, and the growth of mountains.

This normal fault near Lima, Peru, became uncovered over many years, allowing scientists to study it and learn more about fault lines.

In some faults, one section of land is above the fault line. This is called the hanging wall. The other section of land below the fault line is called the foot wall. A normal fault occurs when the two plates move away from each other, causing the hanging wall to slide downward and the foot wall to slide up. In a reverse fault, the two plates are coming together. This causes the hanging wall to slide up and the foot wall to slide down.

The third type of fault is the strike-slip fault. In this case, there is no hanging wall or foot wall. Instead, the two sections of land are level with each other. They don't move toward or away from each other, rather they slide past each other. This action causes seismic activity, but usually not volcanic activity. Strike-slip faults don't create mountains. This is because they aren't drifting apart, which allows lava to form new land. They aren't pushing together, which causes one plate to rise into the air.

Over the years, pressure builds up between the plates forming a strike-slip fault as they try to move. In time the pressure becomes too much. It's suddenly released and the plates settle into new positions. This causes earthquakes.

STRIKE-SLIP FAULT

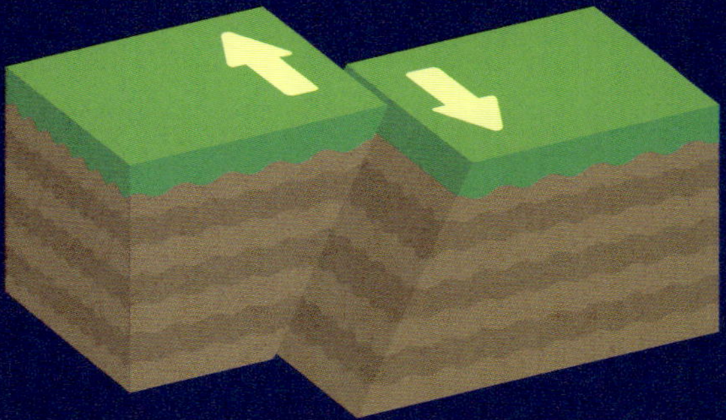

EARTH-SHAKING FACT!

The Alpine Fault forms part of the border between the Pacific plate and the Australian plate. Scientists have discovered that the area has experienced four large earthquakes in the past 900 years—and there will likely be more in the future.

The Alpine Fault is a strike-slip fault that runs through the South Island of New Zealand. It's easy to see from space. The Southern Alps, a mountain range, runs along the edge of the fault on the Pacific plate.

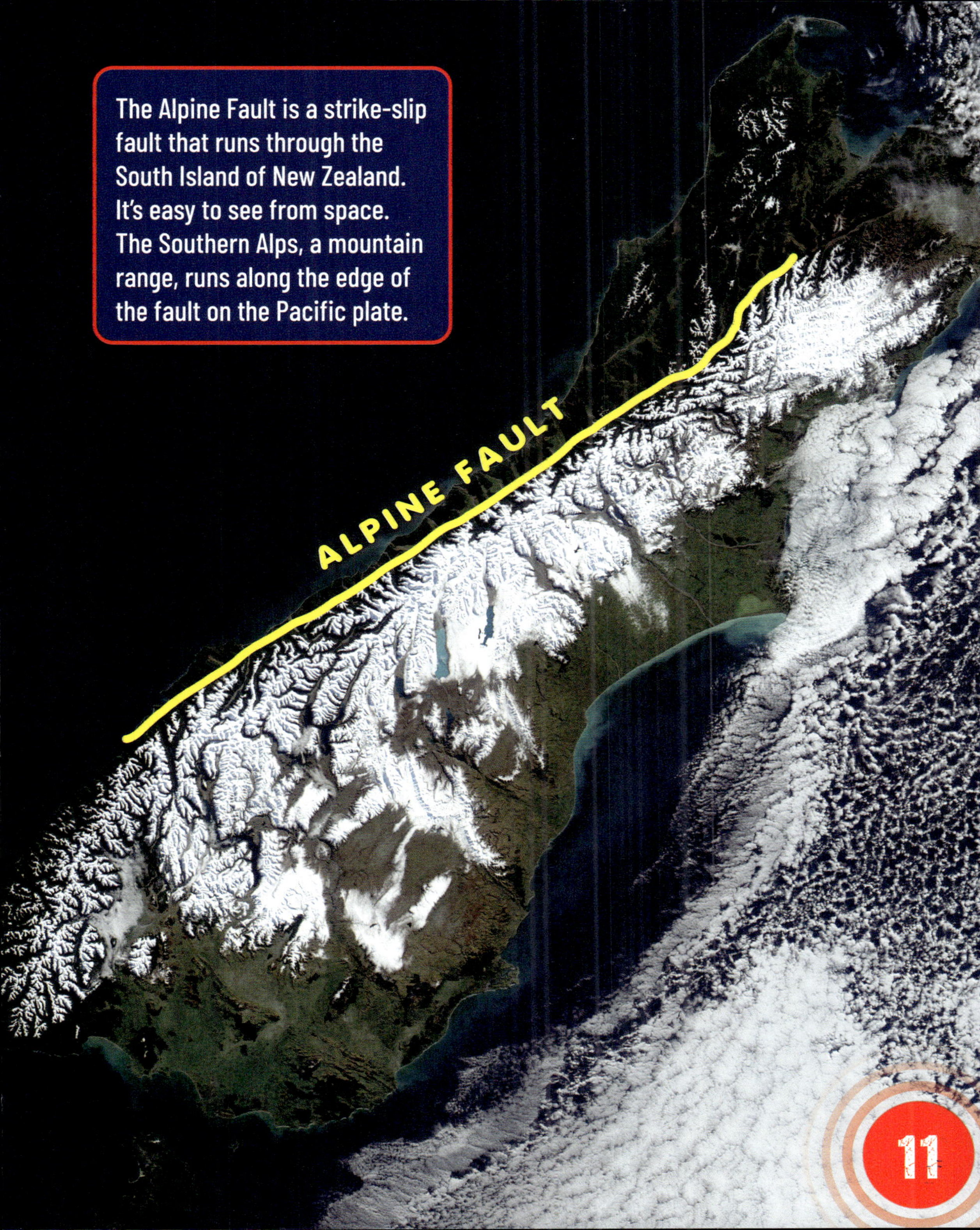

SAN ANDREAS FAULT

The San Andreas Fault is a strike-slip fault between the North American plate and the Pacific plate. The fault stretches for about 800 miles (1,290 km). It starts south of the Salton Sea in Southern California. It passes through the city of San Francisco, follows the coast, and turns west into the Pacific Ocean at Cape Mendocino. The fault is at least 15.5 miles (25 km) deep.

The San Andreas fault is the largest and most active fault in California, but it's not the only one. **Geologists** in California say there are more than 15,000 known faults in the state, and there's probably more. All these fault lines tell scientists a great deal about the landscape of California, it's mountain ranges, and of course, it's history of seismic activity.

Hundreds of years ago, the Salton Sea held far more water than it does today, and the area experienced far more earthquakes. Scientists believe the heavier load of water caused greater seismic activity.

ANDREW LAWSON

Andrew Lawson was a professor of geology at UC Berkeley in 1895. That is the year he discovered San Andreas Fault, which he called a rift, on the Bay Peninsula of San Francisco. Lawson named the fault after Lake San Andreas, through which it passes. In the days after the earthquake of 1906, Lawson formed a team of geologists who studied the destruction and changes to the California landscape. Their findings appeared in the influential Lawson Report.

The Pacific plate and the North American plate are both moving northeast. However, the Pacific plate is moving faster than the North American plate. This means that the movement of the North American plate, **relative** to the Pacific plate at the San Andreas fault, is southwest. This relative movement is called slip rate.

San Andreas's average slip rate since the early 1900s has been about 1.6 to 2.4 inches (4 to 6 cm) a year. During the 1906 earthquake, some sections of the fault moved as much as 21 feet (6.4 m)! At the rate at which the plates are moving by each other, scientist say the cities of San Francisco and Los Angeles will some day be next to each other!

EARTH-SHAKING FACT!

The slow but constant movement around the San Andreas Fault results in many small, daily earthquakes, as well as an occasional large, destructive earthquake. In 1857, a powerful quake struck the Carrizo Plains in Southern California. San Francisco had experienced earthquakes in 1864, 1898, and 1900. This activity had many people worried about future earthquakes.

The San Andreas fault is clearly visible in parts of California. The section of the fault shown here is located in the Carrizo Plains about 100 miles (161 km) northwest of Los Angeles, California.

THE 1906 EARTHQUAKE

The deadliest earthquake in U.S. history occurred near San Francisco on April 18, 1906. Around 5:12 in the morning, a **foreshock** occurred. About a half minute later the main quake occurred, lasting about 42 seconds. Although there were no scientific scales to measure the power of the earthquake at the time, scientists today estimate that it was between 7.7 and 8.3 on the moment **magnitude** scale.

Eyewitness reports described a loud roar during the quake. Many people were just getting up for the day, and many others were still in bed. Plates crashed, walls cracked, and floors shifted! People fled their crumbling homes. Immediately after the quake, fires raged through the city, continuing for several days. When all was over, more than 3,000 people had died and around 250,000 people were homeless. More than 28,000 buildings were destroyed.

The writer Jack London and his wife lived in Glen Ellen about 50 miles (80 km) away from San Francisco. They rode horses into the city. London detailed the destruction for *Collier's* magazine in May 1906.

EMMA M. BURKE: EYEWITNESS

The following text is from an article from *Overlook Magazine* dated June 2, 1906: "It grew constantly worse, the noise deafening; the crash of dishes, falling pictures, the rattle of the flat tin roof, bookcases being overturned, the piano hurled across the parlor, the groaning and straining of the building itself, broken glass and falling plaster, made such a roar that no one noise could be distinguished."

As the quake came to an end, the city fell into **mayhem**. People ran out of their homes as the walls fell down around them, although many didn't make it out. Chimneys and walls tumbled into the streets, injuring people and animals passing by. Cattle stampeded through the streets. Wood and coal stoves around the city fell over and started fires.

Police officers and firefighters quickly sprang into action, but in many cases there was little they could do. Broken water pipes meant hoses were useless. The fires soon swept across the city. Firefighters were ordered to block one quickly moving blaze by blowing up buildings in its path, but this caused more fires.

MOMENT MAGNITUDE SCALE

CATEGORY	MAGNITUDE	DESCRIPTION
micro	less than 2.0	Not felt
minor	2.0 to 2.9	Sometimes felt
light	3.0 to 3.9	Often felt
moderate	4.0 to 4.9	Noticeable shaking, significant damage unlikely
strong	5.0 to 5.9	Can cause significant damage to buildings
major	6.0 to 6.9	Significant damage in populated areas
great	7.0 and higher	Serious damage

As fires burned for days, there was little people could do but watch from a safe distance.

HAM AND EGGS

One of the most widely known events from the morning of the earthquake is called the "Ham and Eggs" fire. A few hours after the quake struck, a family living on Hayes Street began to make breakfast. However, they were unaware that their chimney had collapsed. When they lit the stove the kitchen exploded. The fire would be one of the most **devastating** in the coming days, resulting in the destruction of City Hall, the famous Palace Hotel, and much more.

Fires burned out of control for four days. Federal soldiers soon arrived and did what they could to help. They put out fires and stopped **looting** from damaged buildings. However, nearby Chinatown was largely ignored, and looters, including many soldiers, stole as much as they could.

About half of the city's population (200,000 people) were homeless. They weren't able to cook inside for fear of sparking new fires. Some cooked meals in the streets. Other stood in long food lines. Many of them set up tents in parks. Soldiers set up 11 "relief house" sites that rented tiny cottages to thousands of people while rebuilding took place. **Aftershocks** in the following days continued to cause minor destruction and kept people in a constant state of worry.

It took more than a year to clean up the city of San Francisco after the earthquake of 1906.

MAYOR EUGENE SCHMITZ

A popular musician and labor leader in San Francisco, Eugene Schmitz was elected mayor of the city in 1901. During the quake, Schmitz quickly formed a safety response group of local business leaders, newspapermen, and politicians called the Committee of Fifty. Schmitz made the illegal order for soldiers to shoot looters on sight. As a result, more than 500 people, some innocent of any crime, were shot and killed. Some tried to bring charges against Schmitz, but he was never sentenced.

THE WORLD SERIES QUAKE

Another powerful earthquake struck the San Francisco Bay area on October 17, 1989. The quake lasted 15 seconds and had a magnitude of 6.9. Although it wasn't as strong as the 1906 quake, it caused significant damage in San Francisco and nearby Oakland. Scientists use two names when referring to the 1989 earthquake: San Francisco-Oakland earthquake and Loma Prieta earthquake (the name of a mountain peak where the quake struck). However, it's often referred to as the World Series earthquake because it interrupted game 3 of the 1989 World Series!

The 1989 quake was caused by a fault in the San Andreas Fault system. This includes San Andreas, but also many other smaller fractures near it or connected to it.

The fault that caused the 1989 earthquake is known as an **oblique**-slip fault. This is a fault that acts somewhat like a dip-slip fault and a strike-slip fault.

THE EARTHQUAKE SERIES

The Loma Prieta quake occurred at 5:04 p.m., just 26 minutes before the start of Game 3 of the World Series at Candlestick Park in San Francisco. The two teams in the World Series were both local teams—the San Francisco Giants and the Oakland Athletics. The earthquake began as announcers Al Michaels and Tim McCarver discussed the game. Michaels was heard saying, "I'll tell you what, we're having an earth—" before the TV signal went black. The game was cancelled for the day, but the Giants went on to win four games to none.

The Loma Prieta earthquake resulted in 63 deaths and nearly 3,800 injuries. The San Francisco marina was hit hard. Part of it had been built on filled-in ground. The loose sand experienced **liquefaction**. Other areas experienced landslides.

This section of the Nimitz Freeway had two levels of roadway. The top level collapsed on the lower level, trapping many people.

SAN FRANCISCO–OAKLAND BAY BRIDGE DAMAGE

Property damage was estimated to be at least $6 billion. More than 950 houses were destroyed and many more were damaged. The quake damaged transportation systems in and around the Bay Area. Most deaths resulted from collapse on the Nimitz Freeway. Part of the top deck of the San Francisco–Oakland Bay Bridge collapsed into San Francisco Bay. City officials established a temporary ferry service to make up for the loss of freeways.

A few minutes after the quake, a 5.2-magnitude aftershock struck. There were thousands of smaller aftershocks in the week that followed.

LEARNING FROM THE PAST

California's most populous areas are positioned along or near the San Andreas Fault. Future earthquakes have the potential to affect the lives of millions of people. The San Francisco Bay Area Rapid Transit System (BART) includes a tunnel that was built straight through the fault zone! It seems as if there's not much that can be done to protect people from large quakes.

Earthquake preparedness, safety, and response are important topics for residents and leaders in California. After the 1989 earthquake, all bridges in the Bay area were made more durable to seismic activity. A system of seismic observatories has been established across the state to improve reactions to quakes. The U.S. Geological Survey (USGS) has set up an early warning system that detects earthquakes before people feel the tremors.

This is one of many earthquake observatories in California. This station detects and reports seismic activity and helps people respond quickly to potentially dangerous earthquakes.

EARTH-SHAKING FACT!

After a destructive earthquake hit the San Bernadino area in 1971, California officials passed the Strong Motion Instrumentation Program (SMIP), which established an early warning system of seismic observatories across the state. The Alquist-Priolo Earthquake Fault Zoning Act states that no new structures can be built within 50 feet (15.2 m) of an active fault zone (of which California has 11,000).

A MATTER OF TIME

In a single 24-hour period, California can experience dozens, if not hundreds of minor quakes. Most aren't event felt! However, larger quakes happen several times a year, and destructive quakes aren't uncommon. On July 4, 2019, a 6.4-magnitude quake occurred near the city of Ridgecrest, followed by a 7.1-magnitude quake later in the day. On December 5, 2024, a 7.1-magnitude earthquake hit off the coast near Eureka, California, followed by 216 aftershocks. A **tsunami** warning was issued as well.

What do all these quakes mean? Do they tell a tale of more destructive earthquakes to come? Geologists say yes! The San Andreas Fault is just one of many faults crossing California, and movement in one can cause movement in others. Many scientists believe it's only a matter of time before the next big one hits California.

Seismic activity on and around the San Andreas Fault will continue to shape the land of California for many years to come.

GLOSSARY

aftershock: A smaller earthquake that follows the main shock of a larger earthquake.

controversial: Likely to give rise to disagreement.

devastating: Causing great harm or damage.

foreshock: A smaller earthquake that comes before the main shock of a larger earthquake.

geologist: A scientist who studies rocks and the earth.

liquefaction: An effect that takes place during an earthquake that causes sandy, wet soil to act like a fluid.

loot: To steal something by force, often in times of war or disaster.

magnitude: Size or extent.

mayhem: Violent disorder.

oblique: Having a slanting direction or position.

relative: Existing in comparison to something else.

significant: A noticeably large amount.

tsunami: A great sea wave produced especially by an earthquake or volcanic eruption under the sea.